Book 1: Natural Remedies For Beginners

Book 2: *Organic, Natural Antibiotics And Antivirals For Beginners*

By Dr Alex Nelson

Natural Remedies For Beginners

Disclaimer

Contents

Introduction

Chapter 1: Natural Remedies For Mild Infections

Chapter 2: Natural Remedies For Mental & Neurological Conditions

Chapter 3: Natural Remedies For Gastrointestinal Conditions

Introduction

Every species that inhabits the planet Earth can be considered very lucky. For starters, we have been given the opportunity to exist in a planet that is actually habitable, unlike other planets which are either too cold or too hot for living creatures to survive. We have also been endowed with a very rich environment, which is filled with natural resources and provisions from which both plants and animals could benefit. In short, the solution to all of our problems can be found in the very environment that we inhabit and everything that we could ever ask for has already been given to us by nature. Everything, in this context, consists of but is not limited to food, water, clothes, shelter, , protection, wealth, occupations, fun and recreation, love, and even cures to the diseases and illnesses which we acquire from time to time.

Speaking of natural cures and remedies, it is still a great mystery as to why a lot of people are still not open to the idea of using cures that come from plants and herbs, and instead prefer synthetically produced drugs over these natural cures. Apart from the fact that these synthetic drugs are expensive, they can also introduce toxins to our bodies. However, we still patronize these drugs and continually utilize them. We have already become a society that's dependent on synthetic drugs and is blind to the side-effects of these drugs.

One probable explanation for our neglect of natural cures is ignorance. We lack the necessary information about the benefits of using natural remedies and what the cures are,which is why we don't have faith in them. We often doubt the effectiveness and legitimacy of these remedies because we are accustomed to using synthetic drugs that are produced by multinational pharmaceuticals.

This ebook aims to open a new door for you, a door

towards the appreciation of the natural remedies that even our ancestors have used since the beginning of history. This ebook also showcases the different natural cures and remedies that can aid the common ailments and diseases that we acquire in order to educate you about the healing powers of nature. The majority of these cures are highly accessible and can be easily purchased in the market or can even be found in your gardens. Some are just simple tips and procedures that you can easily follow to prevent acquiring some common diseases.

The main topic that will be talked about in Chapter 1 is the natural remedies that can cure mild infections like cough, cold, and sore throat.

Chapter 2, on the other hand, will tackle a more sophisticated set of conditions. It will talk about the natural cures to mental and neurological conditions like anxiety, depression, vertigo, and insomnia.

The natural remedy for gastrointestinal conditions is the main concept addressed in Chapter 3.

Chapter 4 will look into the ailments that affect the skin and other external conditions.

Chapter 5 will tell you how to battle daily ailments, the natural way.

And lastly, chapter 6 will give you skin care and health tips that you can follow in order to better take care of your children's health and well-being.

Chapter 1

Natural Remedies For Mild Infections

Among the most common infections that humans acquire every year include colds, cough, flu, and sore throat. These are the most prevalent reasons why people take an absence from jobs or school and are the main illnesses that employees use to explain why they are calling in sick from their responsibilities at work. The worst thing about these infections is that they don't choose the target. Anyone who has a weak immune system can be hit, no matter how young or old, or how poor or wealthy these people are.

Although these are just mild infections and are, almost often, not life threatening, they still cause us to feel tired and down. When these infections hit us, our bodies find it hard to function properly and perform our daily routines, and these can be great sources of annoyance and burden to most of us.

Now, you might be wondering: What causes these infections? How can they be cured or if not, prevented? What can we do to minimize the discomfort caused by these infections? This chapter will address all of those questions and teach you how to remedy these infections with the help of Mother Nature.

Common Cold

Colds are one of the leading illnesses that people acquire in every passing year. It is a common infection for the reason that there are a lot of viruses (almost 200 different strands and kinds of virus) that cause this ailment. The most popular virus that can cause colds is the rhinovirus and it accounts for at least 10 to 40 percent of the total number of cases of colds.

Getting a cold may not seem serious but it can cause you a great deal of discomfort and inconvenience, especially if you have important things to attend such as a major exam or a presentation of a big business proposal. It is not easy to think and do your chores properly if you have to constantly pause for a moment in order to sneeze or blow your nose.

Start and symptoms of a cold

Having a cold begins at the moment when you acquire the virus and your immune system fails to get rid of it. Once the virus overcomes the power of your immune systems, the cold virus then attaches itself to the linings of your nose or throat. It then confuses your cells to produce copies of the virus, thus causing them to multiply.

Usually, the symptoms to watch out for in order to determine that you are a victim of the cold virus are repeated and abnormal sneezing, an itchy or drippy nose, sore throat, congestion in the nasal area, watery and teary eyes, and or a fever and even muscles aches.

How can someone catch a cold?

So what can expose you to the possibility of catching a cold? There are multiple of causes for people to catch this virus. Firstly, the cold is the kind of infection that is highly contagious and spreads easily – that means being around someone or interacting with someone who has the cold virus can also make you vulnerable to the infection. The virus can be passed on through shaking a person's hand, kissing them, talking to them, or even just being around them when they unfortunately happened to sneeze. In other words, the cold virus can be passed on through body fluids.

There are also other precursors that can let you catch a cold. Some of these are fatigue and stress. Being constantly exposed to stressful environments or to people who constantly make you stressed and tired can eventually lead to a weakened immune system. Stress can cause your built-in defences to crumble and when this happens, you cannot protect yourself from harmful bacteria and virus. These microorganisms can easily invade your body and cause havoc in your bodily systems.

The weather condition can also cause someone to have a cold. These days, the weather is somewhat moody. One moment it's sunny and then after a few hours, it suddenly rains. You just cannot predict the weather and when you are not ready for the instant change, you become prone to infections such as colds.

Another reason for which people catch colds is when they become exposed to things that trigger allergies. Once your allergies start to act up, colds usually follow.

Remedies

Most colds only last for 10 days. However, we do not want to suffer that long, do we? Here are some of the natural procedures that you can try in order to relieve the inconvenience that the cold virus can bring:

- **Rest**
  - When colds hit you, the best thing that you can do is rest. Yes. That means you have to take a little break from your work or from school. Further stressing and working up yourself will not do your immune system any good. You need to recharge your energy and allow your body to heal itself in order to battle the infection. Also, you will not be the only one who can benefit from your absence. You will also avoid passing the virus to others by staying in your bed and resting. Just imagine the

discomfort if you really yourself force to take that Statistics test with a drippy nose or if you are sneezing every five seconds in front of your boss. So, stay warm and comfortable in your bed to give yourself the chance to get better.

- ### Drink lots and lots of water
  - The rule of thumb in treating a cold is to always rehydrate. When you sneeze and blow through your nose, you also lose body fluids and water. If you don't drink a lot of water, your throat and nose will become more irritated. Don't allow that to happen.

  - Drinking lots of water can also cause the mucus to become easier to flush out from your body, thus making it easier for your body to get rid of the virus.

- ### Steam it up
  - When you have nasal congestion whenever you have a cold, it really is very hard to breathe. You can remedy this by inhaling some steam. The steam can ease the passage of air through your nostrils and passageways, helping you breathe more easily.

  - Like drinking lots of water, inhaling some steam can also help the mucus or the phlegm become softer and easier to blow out of your nose.

- ### Let it go
  - One of the basic ways on how to get rid of your cold is to blow it out from your body. Don't keep

it in your body, as it is not meant to be there. If you do keep it from coming out, the virus will spread and multiply in your body. However, it is also imperative that you know how to blow your nose properly. Aggressive blowing can hurt your nose and it can get scratched in the process. When blowing your nose, you have to use soft tissue. Always avoid the rough ones. Also, make sure that you blow your nose gently so you don't strain the linings of your nose.

Food to eat

When you have a cold, you must make sure that you overload yourself with fruits that are rich in Vitamin C. You need Vitamin C in order to help your immune system fight the virus that causes the infection. So, it would be helpful if you munch on fruits like lemons and oranges. Hot soup can also help, especially if you have nasal congestion. Please avoid cold water.

Omega-3 fatty acids have the power to lessen inflammations. So, if you have a cold, you can eat fish rich in omega-3 fatty acids like tuna. Also, you can eat some oysters because they contain zinc which battles against viruses that cause colds.

Cough

Coughing is a natural reflex that all of us do whenever there is something that irritates our throat. It is our body's way of eliminating foreign substances that cause irritation. However, coughing is not normal anymore when phlegm is already involved. The presence of phlegm when you cough already means that you are infected with a virus or bacteria.

There are generally two kinds of cough: acute and chronic cough. Acute cough only lasts for more or less 3 weeks while chronic cough can go for eight weeks or more.

Causes

Cough can be caused by a lot of factors. Like the cold, it is commonly caused by a respiratory tract infection brought about by a virus. However, a virus does not cause some cases of coughing, like in pneumonia. Instead, the culprits can be bacteria. Sometimes, cold, cough, and flu occur together and can make someone's life miserable.

The virus or bacteria that can cause cough can also be easily passed on through body fluids so sometimes you can catch it from someone who has it. Other causes of coughs include smoking or inhaling smoke from cigarettes and asthma attacks. Pollution from the environment or dusts can also cause someone to have a cough.

Things to avoid

It is a common adage that *Prevention is always better than cure.* So, as much as possible, try to avoid people who have the virus or, if it you can't help it, maintain minimum contact with them. Always wash your hands after holding or interacting with these people and avoid putting your hands on your mouth. Also avoid sharing utensils with people who have a cough in order to prevent the infection from spreading.

Another thing that you can do in order to make yourself invincible from the virus or bacteria that cause cough is to keep your body healthy. Strengthen your immune system by eating healthy foods, having enough sleep, and exercising regularly.

Remedies

Cough? Mother Nature has an answer to that problem too. There are a lot of herbs and plants that can cure common colds and the best news is that the majority of these plants can be easily found in your home or in your garden. Here are some of the plants that can help you get rid of that terrible cough:

- ***Eucalyptus***

    - Eucalyptus has
      evergreen leaves
      that are naturally
      aromatic. Its leaves
      can be boiled or
      powdered in order
      to extract the oil,
      which can be used for medicine.

    - The minty taste of its leaves can help relieve
      throat irritation.

    - Eucalyptus can also be added to rubs for extra
      cooling effect that can ease difficulty in breathing.

- ***Oregano***
    - Our ancestors since
      time immemorial
      have been using
      oregano leaves in
      order to cure and
      relieve the
      symptoms of cough.

    - You can soak
      oregano leaves in a
      cup of boiling water as you would for a bag of tea
      for a couple of minutes and drink it in order to
      soften dry cough.

- ***Thyme***
  - The Germans have been using thyme as a cure for cough for years.

  - Thyme is an herb that contains flavonoids, which relax the muscles in the trachea and can cure inflammation.

  - You can add crushed thyme leaves into water and infuse them for 10 minutes. After it's ready, strain the water and drink.

- ***Lemon***
  - Lemon is known for its antiseptic and anti-inflammatory properties.

  - It is also a fruit rich in Vitamin C, which helps in fighting colds and cough.

  - To ease throat irritation, you can suck on a quarter of lemon. If it is too sour for you, you can add a little bit of salt to it.

Aside from plants, there are also some practices that can save you from the turmoil of having a cough. For example, you can drink a lot of water to keep your throat from drying up and to make phlegm easier to expel. Rest is very much needed by people who get infected by cough too, so they can recharge and replenish their energy. Sleeping with some extra pillows and elevating your head a little bit can aid in the better passage of air through your lungs to help you breathe smoothly and properly.

Foods to avoid

Stay away from mangos and bananas when you have cough because these fibrous fruits can further irritate your throat. Also avoid milk because it can cause the production of excess mucus, making it stickier and thicker. If you get a cough from asthma, please avoid eggs, nuts, and shellfish like shrimps because these can trigger asthma.

Sore Throat

Just like cold and cough, sore throat is also caused by viral infections. Specifically, it is an infection in the voice box caused by a virus. Its' common symptoms include irritation, pain, or itchiness of the throat, or sometimes a combination of the three. A person who has sore throat may experience difficulty in swallowing, especially crunchy and hard chunks of food. One can also have dry throat, white patches on his or her tonsils, and sometimes, swollen neck glands.

It commonly attacks smokers, kids, and, of course, people with weak immune systems.

Causes

The main cause of sore throat is a virus. However, bacteria and having an unhealthy lifestyle can also be factors. People who live in polluted places or those who consume lots of sweets but don't drink lots of water can be more prone to acquiring sore throat.

Remedies

In curing sore throat, there's no need for expensive gargles and medicines. All you have to do is to mix a pinch of salt with warm water then gargle the mixture for a couple of minutes. You can also relieve the pain in your throat by avoiding cold beverages and drinking warm fluids, like soup, instead, .

First Aid Ointments

There are also natural ointments that can substitute the expensive and synthetic ones in the market. Aside from the fact that these ointments are budget-friendly and can easily be found your surroundings, these are healthy for the body as well. One of the most popular natural ointments is aloe vera, which is a very good cure for burns. The gel that comes from aloe vera has a cooling effect on first-degree burns. Another natural ointment is coconut oil, which can be used as a moisturizer for dry skin and is a cure for boils and wounds. Also, you can use honey for wounds, stings, and minor cuts. Honey is a great antimicrobial and it can help wounds heal faster.

Chapter 2

Natural Remedies For Mental & Neurological Conditions

It is not only minor infections that nature can heal. The environment is so powerful that it extends its healing effects to problems that include our mental and neurological well-being. Here are some tips and cures that will help you battle common emotional and psychological problems without the help of a therapist or a doctor.

Anxiety

Anxiety is an ancient response to harmful stimuli. Our ancestors were constantly surrounded with animals that could hurt or kill them. They had to always be alert in order to escape danger. That response has been passed on to modern men even if our environment is relatively less dangerous than before. Being anxious still benefits us because feeling so heightens our defense mechanisms at times of trouble. But, if we get attacked with anxiety almost regularly, it becomes stressful and unhealthy for us. We become jittery and uneasy while our muscles become tensed.

Although fear and anxiety are both responses related to the "fight or flight" system, fear is different from anxiety in terms of the perception of threat. You feel fear when the threatening stimulus is in front of you but it is anxiety that you feel when you think about a threatening stimulus that can strike you in the future.

How to cure anxiety

The secret to minimize the feeling of being anxious is

to always prepare for the future. Sometimes, we become nervous of something because we don't prepare for the possible outcomes that these things can bring. One example is cramming for a major examination. You feel anxious because of the possibility that you might flunk in that examination or it could be failing a job interview or not receiving a promotion. There's a possibility of failure because you didn't study in advance or perform well in the interview. So, for these kinds of circumstances, you can avoid feeling anxious if you plan ahead and prepare for the upcoming stimulus.

If you can't help but feel tensed, you can also have a cup of chamomile tea in order to calm yourself. Chamomile has chemical compounds that can bind together with some receptors in the brain and tell your brain to keep calm and composed. You can also sniff the scent of lavender because it has been proven that doing so can relax tensed muscles and lower your blood pressure to a normal level. Try to eat bananas as well. Bananas have a chemical called tryptophan, which is a precursor to the synthesis of the neurotransmitter called serotonin which can help you feel tranquil and be at peace.

You can also try simple breathing exercises like inhaling and exhaling in order to regulate your breathing. Doing so can also ease muscle tensions and help you feel more relaxed.

Mild Depression

There are just days when we feel so down. This is perfectly normal. We sometimes feel down for no apparent reason because our body cannot sustain being too happy for a long time. Feeling happy all the time is just too energy consuming both for our bodies and our brains. However, prolonged sadness can be very bad for our well-being too.

Usually, it takes more than six months for a person to be declared depressed and when that happens, he or she should seek professional help.

How to cure mild depression

Not all forms of depression need to be treated with synthetic drugs. There are natural tips that you can follow to chase the blues away. For example, thinking of happy thoughts. The brain is so powerful that it can dictate your whole body how to behave. You can feel happy by thinking of happy thoughts because doing so can beget more happy thoughts. Avoid people or things that can make you sad and focus on things that make you laugh or smile instead, like a funny book, a comedy show, or a happy classmate. Do not wallow in your sadness and isolate yourself. Doing so will certainly not do you any good and will further drive you to depression.

Another thing that you can do is to perform physical activities as much as possible. Exercising can release certain neurotransmitters in the brain that can help you chase the blues away. Go outside and take a jog with a friend or play some Frisbee. Go walk your dog or take a stroll in the park. You can feel relaxed and one with nature and feel a sense of fulfilment by doing so.

If you feel sad most of the time, make it a habit to let it out. Don't keep all the hurt for yourself. Tell your mom and talk to her about what's bothering you; you can even tell a trusted friend or your pet. Sometimes, people succumb to their depression because they keep all the negativity in and don't have any outlet where they can channel all their sadness.

Insomnia

Insomnia is the inability to sleep soundly and has three different types. A person can have the difficulty to fall fast asleep, have the inability to maintain a sound sleep (waking up unintentionally from time to time), or have the difficulty to go back to sleep after waking up unintentionally.

How to cure insomnia

One of the causes of insomnia is a messed up sleeping pattern so the basic thing that you can do to prevent it is to establish a proper body clock. Make it a habit to sleep and wake up at a specific time. Avoid staying up late or beyond your usual sleeping hours because doing so can confuse your brain and your body when they really want to rest and fall asleep.

Another reason for insomnia is overthinking. Before you go to bed, avoid thinking about mind-boggling mathematical problems, personal problems, or even the mysteries of the earth. Stimulating your brain can increase brain activity, which will lead you to feel energetic. If you have insomnia, try reciting a memorized prayer over and over. Make sure that you don't think too much in order to calm your brain and order your system to fall asleep.

Drinking chamomile tea and taking a hot bath before going to bed can also help you get a good night's sleep. You will feel more relaxed and sleepy afterwards.

Vertigo

People who suffer from vertigo can feel dizzy and, sometimes nauseous. It is as if their environment is constantly spinning or whirling, causing them to lose balance even if they are just standing or sitting still. One of the common causes of vertigo is a damage or infection in the vestibular area of the ear, which give us our sense of balance. When this happens, we lose our composure and feel dizzy or nauseated because of the whirling sensation that we constantly feel. Other causes of vertigo can include eye problems, brain tumor, head injuries and heart problems, dehydration, and anemia.

How to cure vertigo

First and foremost, you need to have enough sleep because skipping on much needed sleep can lead to the symptoms of vertigo just like dizziness. Having enough sleep can also keep you from being anaemic.

If you ever suddenly feel dizzy, make sure that you sit or lie down at once to avoid falling down and hitting your head on the floor. After doing such, try to rest your head between your legs in order to facilitate faster blood flow through the brain.

Since vertigo is also caused by dehydration, you must make sure that you constantly keep your body hydrated by drinking lots of fluids. Juices, soups, and energy drinks can be helpful as well.

Another thing that you can do to cure your vertigo is to stand up or move gently. Abrupt body movements can only exacerbate the feeling of being dizzy and being sick. You can also try inhaling some minty rubs or liniments that have the aroma of menthol. This can help you feel less sick and less dizzy.

Chapter 3

Natural Remedies For Gastrointestinal Conditions

We all love to eat and eating is one of the simple joys in which almost all human beings indulge. We love to eat comfort food whenever we feel sad, lonely, or broken hearted. Food is even a requisite in any social gathering or activity. Eating is a major part of our day-to-day lives since we acquire the necessary nutrients and minerals that our body needs through the consumption of food. However, food consumption becomes less enjoyable when problems that involve our digestive system hit us. We don't get to appreciate our delicious food well whenever we have these kinds of problems.

We don't have to worry though because surprisingly, there are also natural remedies and cures for gastrointestinal problems. The healing effects of nature do not only cover problems that involve the brain and mild infections caused by bacteria and viruses but it also have available solution for our tummy problems.

Diarrhea

Diarrhoea happens when the large intestines fail to extract fluids or water, which leads to a watery stool. When someone has diarrhoea, he or she frequently goes to the bathroom and his or her faeces are accompanied by lots of water. Bacteria, viruses, or food poisoning can cause it.

Remedies

When you suffer from diarrhoea, make sure that you frequently rehydrate by drinking lots of water to prevent the loss of a lot of electrolytes. You can also eat yogurt in order to increase the number of healthy bacteria that will facilitate in good digestion. Also munch on bananas, toast, and rice in order to soothe your upset stomach. Bananas contain a soluble fibre that can slow down the passage of stool. This fibre can also be found in carrots so you can also munch on those if you prefer.

Constipation

Constipation is characterized by difficult and less frequent bowel movements. Someone who is suffering this can experience pain and difficulty in removing bowels because of hardened stool, which was caused by too much absorption of water by the colon.

The common causes of constipation include low fibre content in one's diet, inactivity, not drinking enough water, and impeding bowel movement.

Remedies

The main solution to constipation is to eat lots of fibrous fruits and green leafy vegetables like papaya, mangoes, and moringa olifera. Fibre is the natural sweeper in our colon and it helps in the faster digestion of our food. Yogurt is also a very good food to eat whenever you have constipation because it promotes the growth of good bacteria in the colon, which are also helpful in digestion. Aside from water, you can also opt for prune juice, which is rich in fibre and contains sorbitol, which softens stool. Drinking lemon juice can also be helpful since its citric acid component stimulates the digestive system.

Bloating & Gas

It is natural for us to pass gas through flatulence and burping. They our body's ways of getting rid of swallowed air and telling us that the food we've consumed have already been broken down. But, there are times when we can't belch or pass gas that's why we become bloated.

Remedies
Eating too many fatty foods and eating too fast can also cause bloating. To prevent bloating, try to eat and chew your food slowly. This allows your tummy to get ready to process that food and doing so will also enable you to enjoy your food better. You should also try to avoid carbonated drinks like sodas because these can increase the amount of air in your stomach and cause bloating.

Nausea
We feel sick or feel like we are about to vomit because of a lot of reasons. Usually, vomiting is a sign that you've been hit by another disease just like the flu or a fever. Vomiting is also our body's way of getting rid of toxic and stale food that we have unconsciously and accidentally ate.

Remedies
Vomiting can also translate to losing water and electrolytes. So, it is only imperative that you replenish lost bodily fluids by drinking lots of water. Another remedy for nausea is eating soda crackers in order to absorb the excess acids in your stomach. Applying pressure on your palm also helps in stopping the feeling of being sick.

Chapter 4

Natural Remedies Skin & External Conditions

Diseases don't only attack us from the inside, but mess up with our skin too. Here are some of the natural ways in which you can prevent common skin problems and conditions.

Sunburn

When we stay too long under the heat of the sun, we become overexposed to its harmful rays, causing our skin to get a little inflamed and burnt.

Remedies

To soothe your inflamed skin, you can try applying aloe vera gel, which is very good in soothing burns, cucumber slices or a cold compress on the burnt areas. These will ease the painful sensation and reduce inflammation. You can also apply oatmeal soaked on cold water or a mixture of cornstarch and water on the burnt area in order to relieve yourself from the painful and stinging sensation.

Avoid using soap for a few days too because soap can further dry up your skin and exacerbate the redness and pain.

Toenail Fungus

This condition is caused by fungus and it causes the nail to become discoloured, broken, or damaged. This is also hard to cure since toenails grow slowly which consequently lead to the bacteria spreading easily.

Remedies

To get rid of this nasty infection, you can soak your feet in apple cider vinegar or apply tea tree oil in the affected are. Apple cider vinegar can avoid the spread of the fungus because of its acidic property while tea tree oil has antiseptic and anti-fungal properties.

Puffy Eyes

Puffy eyes can be caused by the irritation and itchiness in the skin around the eye. A few culprits behind this are allergies, the lack of sleep, or even stress.

Remedies

Drinking lots of water can get rid of puffy eyes by flushing out the excess water that causes some body parts to swell. Tea bags and slices of cucumber are helpful as well, as they soothe swelling and have anti-inflammatory properties and can also help relax the eye muscles.

Rash

Rashes can sometimes occur whenever allergens or bacteria that cause these diseases irritate our skin. The symptoms of rash that you should watch out for are itchy, chapped, and scaly skin.

Remedies

Vitamin C is very important in treating rashes since it brings back the firmness and the moisture of the skin. So, be sure to eat fruits like lemon and oranges because these are known fruits that are abundant in Vitamin C. You can also wash the infected area with chamomile tea, oatmeal, and aloe vera in order to reduce itchiness and in order to moisturize torn skin.

Eczema

Eczema is a painful skin condition that's characterized by rough and inflamed patches on the skin, which are often accompanied by blisters.

Remedies

The oil extracted from sweet almonds can be used as cure for rashes because it contains Vitamin C. Natural skin moisturizers can also be found in coconut oil and aloe vera, which can ease the discomfort of having dry and scaly skin caused by eczema.

Psoriasis

Psoriasis is a chronic skin problem that is characterized by thick, white or red patches on the skin. This is caused by the abnormally fast growth of the skin cells.

Remedies

There are studies that support that claim that avoiding fatty foods, as well as red meat can help cure eczema. Olive oil, coconut oil, aloe vera, and apple cider vinegar can help with the itchiness, the inflammation, and the dryness of the skin because of their anti-inflammatory, anti-fungal, and antiseptic properties.

Chapter 5

Natural Remedies For Daily Ailments

There are also diseases that may not cause us to be bed-ridden but still cause a great deal of inconvenience like heartburn, joint pain, headache, menstrual cramps, yeast infections, and muscle pains. Here are some ways on how you can remedy these problems.

Heartburn

Heartburn is a burning feeling in the chest area caused by too much acidity in your stomach and indigestion. Mixing a spoonful of baking soda in water in order to neutralize the acidic state of your stomach can easily cure this. You can also eat some bananas or apples because they have natural antacids that remedy hyperacidity. Surprisingly, chewing gum is also considered a cure for heartburn because doing so can stimulate the salivary glands to produce more saliva in order to get rid of the excess acid in your stomach.

Joint Pain

Joint pains are also known as soreness in the joints. It is common in aged individuals and people who either had too much exercise or those who don't exercise at all. You can cure the inflammation of your joints by using ginger tea. You can also massage virgin olive oil on your sore muscles in order to relax them and ease the pain. It is also important that you munch on spinach, nuts, and legumes because these foods are rich in magnesium that can help prevent sore muscles and joints. Exercising regularly also helps too because doing so can loosen the joints and enable you to move easier.

Headaches
There are numerous ways on how you can relieve yourself from headache. You can massage your temples in order to release certain neurotransmitters that act as natural painkillers. Sometimes, headaches are caused by stress. In order to cure it, you can try getting out to inhale some fresh air and relax yourself. You can also look or focus on something green like a leaf or a plant since doing so can give you the feeling of being relaxed.

Menstrual Cramps
Menstrual cramps are something that most women can relate to. It is caused by contractions in the uterus and this usually occurs during the onset of menstruation. Its symptoms include pain or pressure in the abdomen.

You can ease the pain by taking a warm bath and doing some exercise like yoga.

Yeast Infection
One the most common vaginal infection that women acquire is the yeast infection. It is caused by the overgrowth of fungus in the vagina. Luckily, this is highly curable. You can eat yogurt in order to help the good bacteria in the vagina grow in number and fight the fungus. You can also rely on garlic for its anti-fungal properties and place a clove of garlic on your lady part while sleeping.

Muscle Pain
Watermelon juice has been proven to treat muscle pains because it contains an amino acid called l-citrulline that fights muscle pain. You can also opt for a warm bath

or a nice massage in order to relax the muscles. You can drink a mixture of some apple cider vinegar and water too for muscle relaxation.

Chapter 6

Natural Health & Skin Care Recipes For Children Under 12

Children have relatively a weaker immune system than adults. They need special care and attention so that they will not easily acquire illnesses and diseases. Their skin is highly sensitive too and prone to skin infections and complications.

Here are some simple tips to protect your child from the threat of infection and diseases.

1. ***Make sure that your child is eating a balanced diet.***

Diet is everything. Eating the right kinds of food can help your child reach his or her full potentials and can protect him or her from being hit with terrible illnesses. Make sure that your kid is eating healthily and keeping a supply of fruits and vegetables. He or she needs to have the three major food groups in every meal too to acquire the essential nutrients and minerals that are needed by the body. Discourage your child from eating junk and fatty foods.

1. ***Let your child exercise regularly.***

Make sure that your child doesn't get too caught up with video games and the Internet and still have them go out and play outside. Go out with your children for a run or a stroll in the park as a boding session.

2. *Teach your child to observe proper hygiene.*

Let your child understand that the key to avoiding diseases is by observing cleanliness in his or her body and in the surroundings. Teach your children the importance of hand washing before and after every meal. Make sure that they take a bath every day and that they change into dry and clean clothes after playing outside to prevent rashes or the spread of bacteria.

Natural Skin Care Recipes

Before applying any natural skin care remedy, you have to research and read up on your ingredients or the herbs that you are going to use. Make sure that you have substantial information about the plants just to be safe.

You can make natural moisturizers by mixing an emulsifier with water and a kind of oil of your choice. Just make sure that your child is not allergic to the oil that you're going to use.

You can also extract coconut oil from coconut juice or water by heating the juice for more or less 1 hour, until the juice becomes crystal clear. Coconut oil is gentle for the skin and is great in treating dryness.

Conclusion.

It is very comforting to know that we have an environment that could provide us with natural cures that don't introduce harmful toxins to our bodies. It is also great to know that some of the common diseases that we suffer from can be treated without having to take synthetic medicines.

Hopefully, this ebook inspired you to try out some of the natural cures that can serve as alternative to synthetic drugs and encouraged you to make your own homemade first aid.

Organic Natural Antibiotics And Antivirals For Beginners

By Dr Alex Nelson

Disclaimer

This book is intended to be a general guide, to raise awareness, and to help people make informed decisions in the context of their own personal circumstance. As everybody's circumstances are different, so are the remedies you should seek. While many of the recommendations in this book can be applied by almost anybody regardless of their conditions they are not intended to and should not be relied upon to replace personal medical advice.

The author accepts no responsibility for any loss or injury, be it personal or financial, as a result for the use or misuse of the information in this book. If you have any doubts or concerns after reading this book, please speak to a doctor or other qualified person before taking any actions.

Contents

Introduction

Introduction

Our bodies are equipped with a built-in protection system, called the immune system that defends us from nasty and health-threatening bacteria and viruses. Our immune system is amazingly crafted giving us the power and the capability to protect ourselves from internal threats, which may not be noticed by our naked eyes. Whenever we get hit by bad bacteria and harmful viruses our immune system acts either as our army and attacks these foreign invaders or as some kind of wall and stop them from spreading throughout our bodies.

However, our protection wall and our army are not always strong enough to battle the harmful bacteria and viruses. Some of them are too sophisticated for our built-in protection to handle. Also sometimes, because of our unhealthy lifestyle and the amount of stressors that we are exposed to every single day, our immune systems and resistance to pathogens go down and become weak to fight these bacteria and viruses. When these circumstances happen, we already need to turn to the aid of antibiotics and antivirals to back our immune systems up and help fight these invaders.

Now, the question is: **What exactly are antibiotics and antivirals and what are their differences or similarities?** This ebook will answer these questions and at the same time discuss relevant topics that are concerned with antibacterial and antivirals and with their uses.

Chapter 1 is an overview about antibiotics and antivirals. The difference between the two will be explained, as well as their specific uses.

Chapter 2 will tackle on the advantages and the disadvantages of using synthetic antibiotics.

The benefits of using natural antibiotics and antivirals will be explained in Chapter 3. You will know the comparative effects of natural substances versus synthetic ones.

Chapters 4, 5 and 6 will enumerate some of the herbs, foods and oils that have natural antimicrobial properties.

And lastly, Chapter 7 will give you tips and recipes on how to craft your own natural antivirals and antibiotics right at the vicinity of your own kitchen.

Chapter 1
An Overview Of Antibiotics And Antivirals

Antibiotics and antivirals help our bodies fight foreign organisms especially when our immune system is too weak to fight these organisms. They serve as back up or substitute fighters at times when our built-in protection cannot fight back.

So, what really are similarities and difference of these so-called antibiotics and antivirals? When can we use an antibiotic or antivirals? This chapter will answer frequently asked questions about them as well as provide an in depth information about their uses and correct your preconceived notions about these substances.

Virus VS Bacteria

Before we discover the similarities and the differences between antivirals and antibiotics, it is imperative that we know first the difference between a virus and a bacterium because the distinction of the two paves the way to the understanding of the difference of antivirals and antibiotics. Identifying whether an infection has been caused by a virus or bacteria is the key to knowing whether which kind of antimicrobial substance is more appropriate to use, antiviral or antibiotics.

Both bacteria and virus often cause infections that have almost the same symptoms, making it difficult to determine which of the two really caused the illness or disease. They also both have the ability to mutate and build resistance against antivirals and antibacterials, making them serious forces to be reckoned with.

However, the difference between bacteria and a virus lies in the absence or presence of a host. Bacteria are microorganisms that are ubiquitous and can adapt to any environment in order for it to survive. Although some bacteria are detrimental to one's health, good bacteria also exist and they are mainly present in the intestines as aid for digestion. Viruses, on the other hand, are smaller microorganisms and depend on a host for its survival. There is a cessation in the growth of viruses when they have no host to feed on. Unlike bacteria, viruses cannot reproduce. Instead, they mess up the mechanism of the cells and dictate them to produce the virus. Compared to bacteria, viruses develop resistance to antimicrobials faster. Bacteria may build resistance after a month or a year but viruses develop resistance and mutate as rapidly as a day or a week after.

Viruses cause diseases such as colds and flu while bacteria cause wound infections.

Antibiotics

The word "**antibiotics**" came from the Greek words "**anti**", which means opposed to or against, and "**bios**", which means life. Basically, "*antibiotics*" means "*against life or opposed to life*". To be more scientific and specific, an antibiotic is a substance that either fights bacterial infections and kills them or prevents the reproduction or growth of the bacteria that are causing the infection and impede them from multiplying. This substance originates from a microorganism, which control the spread of bacteria or exterminate other bacteria.

Antibiotics are also commonly known as antibacterials because their main purpose is to fight infections that are caused by bacteria, not by viruses. Almost all antibiotics are not used to cure infections brought about by viruses and it must be emphasized that antibiotics do not cure and are not effective in treating viral infections.

The History of Antibiotics

    The history of the use of antibiotics can be traced way back in the ancient civilizations. People utilized plants and other natural concoctions to treat their wounds, bites and other infections. For example, moulds from bread were used as treatment for wounds by the ancient people of Greece and the Chinese also utilized different kinds of herbs to treat infections. These just go to show that the use of antibiotics to treat infections is not a recent discovery in the field of medicine.

    Although ancient people already used natural antibiotics, they most probably didn't know the science behind it. The people in the modern age provided the missing information about bacteria and antibiotics that the ancient people failed to give. Modern Age scientists treated bacteria more scientifically and logically and numerous names surfaced and became popular because of their relevant contributions in the field of microbiology and bacteriology. One of these names is the name of Alexander Fleming.

    Sir Alexander Fleming invented the Penicillin, the world's first and most popular antibiotic, in the year 1928. This invention was an honest coincidence and a serendipitous discovery but it was such a great discovery that opened new doors in the field. It was a great stepping-stone to the formulation of modern antibiotics.

Classifications Of Antibiotics

    Antibiotics are commonly classified according to the range of organisms that they can target, its way of administration, and their function.

    In terms of the range of organisms that they can target, a certain kind of antibiotic can be classified as broad-spectrum or narrow-spectrum. Broad-spectrum antibiotics, just like the Fluoroquinolones, only means that these kinds of antibiotics can target a wide range of bacteria and other microorganisms and can treat several kinds of bacterial infections. The narrow-spectrum antibiotics can cure only a lesser number of bacterial infections compared to the broad-spectrum antibiotics.

Antibiotics are also classified according to the way they are taken in. They can be taken orally, meaning they are ingested in the form of capsules or syrups; topically, meaning they are applied as lotions or creams or ointments directly to the affected body parts; or through injection, which means you have to infuse the antibiotics to your systems. Topical antibiotics are usually used as treatment for skin infections while injectable antibiotics are used for more serious bacterial infections.

Lastly, antibiotics are also grouped according to what they can do. Antibiotics can either be bactericidal, meaning their purpose is to kill and impede the bacteria's cell wall from forming; or they could be bacteriostatic which means that these kinds of antibiotics will stop bacteria from growing in numbers. An example of a bactericidal antibiotic is Penicillin and examples of bacteriostatic antibiotics are tetracyclines.

Types Of Antibiotics

Here are some of the common types of antibiotics and their respective uses:

Penicillin
- These are used for a wide variety of bacterial infection including bacterial infections in the urinary tract.

Aminoglycosides

- Aminoglycosides are among the very powerful antibiotics which must be taken carefully and with proper prescription because it can cause serious injury such as damaged kidneys and loss of hearing when taken in high dosage.

- Since this kind of antibiotic breaks down easily, it is given through injections.

Fluoroquinolones

- This antibiotic belongs to the broad-spectrum antibiotics and can cure a wide range if infections.

- An example of this antibiotic is Cipro.

Maerolides

- This is a good alternative to penicillin and can treat infections in the lungs and chest area.

Tetracyclides

- Tetracyclides can be used as treatment for severe acne.

Caphalosporins

- This is effective in the treatment of serious infections such as meningitis.

Common Bacterial Infections

Some of the most common infections that people get from bacteria are salmonella, tuberculosis, syphilis, meningitis, strep throat, urinary tract infections, respiratory infections, boils, pneumonia and acne. Bacterial infections differ in the levels of threat that they inflict on humans and their threat basically depends on the kind of bacteria. We don't need to be too worried about some of the bacterial infections, such as acne or strep throat or boils. However, we must seek immediate treatment should we acquire bacterial infections such as meningitis or tuberculosis or syphilis because these infections can be life threatening.

Antivirals

The virus that targets the body and the one that targets the computer system operate in the same manner. The virus in the body makes us their host and causes a disorientation to our cellular processes in the same way that the virus in our computers feeds on the hardware or the software and mess up the normal process of the computer.

The word "**virus**" comes from the Latin word "**vīrus**" which means poison so the term "**antiviral**" literally means "against poison". By scientific definition, antivirals are medications that can fight infectious diseases caused by viruses. They treat infections either by stopping or slowing down the reproduction of the viruses in our system or by boosting the power of our immune system to fight the virus. The power of antivirals are only limited to inhibiting the growth of the virus and stop it from growing and reproducing. They cannot fully kill or exterminate the virus, unlike what the antibiotics can do.

Antivirals are utilized for the treatment of a relatively small range of organisms and there are different varieties of antiviral treatment for a specific kind of infections. They are also used in two ways, either as vaccines or as treatments to infections. As vaccines, antivirals serve as precaution and they prevent infections from developing. When used as treatment, antivirals can reduce the length of time that the symptoms of the infections are felt by a person. In other words, they lessen the damage that the virus causes.

Common Viral Infections
Like bacterial infections, viral infections are also placed in a continuum, depending on the level of harm that they can cause to our bodies. Some viral infections can be treated at once but some can cause extreme consequences such as death.

The common viral infections that are suffered by millions of people include colds, flu and other sophisticated strains of influenza such as the swine flu or the H1N1, hepatitis, sexually transmitted infections such as HIV or Human Immunodeficiency Virus, AIDS or Acquired Immune Deficiency Syndrome, herpes, and now, the dreaded Ebola virus infection.

Antibiotic VS Antivirals
The definitions of antibiotic and antiviral, as well as their uses have already been tackled in the previous paragraphs. At this juncture, the two will be contrasted and compared side by side in order to foster a better understanding about the two and a clearer distinction between them.

Similarities
Antivirals and antibiotics are similar on three grounds: (a) they both target the microorganism that is causing the infection, (b) they are both kinds of antimicrobials, which mean they can either destruct the microorganism or they can inhibit its growth, (c) and they both can be taken in various ways such as oral, topical or through injection.

Differences

There are clear differences between antibiotics and antimicrobials too even if they are both under the classification of antimicrobial substances. For starters, antibiotics are for bacterial infections while antivirals are for viral infections. Antibiotics cannot also treat infections caused by viruses. Unlike antibacterials that can kill the microorganisms and stop the bacteria from growing, antivirals cannot destroy viruses and they can only inhibit the growth of the virus. Also, antivirals are much more difficult to develop than antibacterials for the reason that killing the virus can also mean damaging the cells of the host. Another difference between the two is that antibiotics can be used to treat a wide range of infections while antivirals can treat more specific infections.

Chapter 2
The Pros And Cons Of Synthetic Antibiotics

There are a lot of aspects of our day-to-day lives that developed together with the innovation of modern technology. One of these aspects is the way we attend to diseases and illnesses. The discovery of modern medical apparatuses paved the way to the discovery of modern drugs and cures for a new generation of bacteria and viruses. Antibiotics, which were formulated in medical laboratories, replaced natural antibiotics and the once popular Penicillin became somewhat obsolete with the formulation of new synthetic antibiotics.

The synthetic antibiotics rose to fame after Penicillin was observed to be a major cause of allergic reaction to a majority of people and after it failed to cure new kinds of bacteria. Synthetic antibiotics, such as the Amoxicillin became the alternative for Penicillin because it is cheaper, more affordable and more effective.

However, not only the advantages of these synthetic drugs were discovered as they became largely available to the market. Hidden disadvantages that were lurking behind the effectiveness of these synthetic drugs were also discovered. This chapter is devoted to the discussion of the pros and cons of these synthetic drugs.

Pros of Synthetic Antibiotics

The reason that synthetic antibiotics still thrive in the market despite their side effects is because they are very effective. Synthetically produced antibiotics can effectively kill bacteria that are causing the infection. They work and diminish the bacteria fast and they do it efficaciously. In times of great need and in times of emergencies, these synthetic antibacterials can really help avert infection and help save a life.

Cons of Synthetic Antibiotics

One of the most serious side effects of synthetic antibiotics is bacterial resistance. This is a phenomenon whereby the bacteria that a specific kind of antibiotic is supposed to cure become resistant or somewhat immune to that kind of antibiotic. The bacteria then mutate and evolve and become stronger than the antibiotic itself. This happens because most people are abusing the use of and misusing antibiotics. Most of us are guilty of not giving our immune system the chance to fight the bacteria and just being dependent upon antibiotics even if the infection is not that serious. Another reason why bacteria become resistant to certain antibiotics is because we sometimes don't finish taking our prescriptions. I know that a lot of people are guilty of this. We have the tendency to stop taking the antibiotic drugs given to us once we feel better even if we were advised to take them longer. There is a reason why antibiotics are prescribed to be taken for a specific number of days and that is to make sure that the drug kills all of the bacteria. Cutting our medication short can foster the growth of a stronger kind of bacteria that are much harder to cure than their previous kind. These kinds of bacteria cause worse infections that, if not treated, are much longer to kill.

Another disadvantage of synthetic antibiotics is that they are toxic and they can introduce toxins to our systems. These toxins that synthetic antibiotics carry with them inside our bodies can cause our immune systems to weaken, causing it to fail in protecting us from the threat of these bacteria and consequently, making us more vulnerable to bacterial infections.

In addition to these side effects, synthetic antibiotics, due to their powerful and effective ability to kill off bacteria, can also accidentally kill off the healthy bacteria that live in our intestine. These innocent bacteria, which cause us wellness than harm and aid in our digestion, get accidentally killed together with the harmful and bad bacteria. This leads to intestinal problems like diarrhoea and indigestion.

Some of the other side effects of antibiotics are kidney stones, vomiting, allergic reactions especially to Penicillin and Cephalosporin, chronic fatigue syndrome and deafness. Yes you read that right. Some antibiotics are also ototoxic drugs, which means that they can cause damage to the cochlea or the vestibular system. High dosages of some powerful antibiotics like the Aminoglycosides can cause damage to the vestibular system. There was even this popular case of a woman who felt like she was perpetually falling because she was given high dosages of Gentamicin, a kind of antibiotic. The constant use of Gentamicin damaged the structures of her inner ear that are responsible for giving her sense of balance.

And the most important disadvantage in using synthetic antibiotics is that it is harmful to the environment. Since they are synthetic, which means they are made out of mixing chemicals, they have the potential to pollute the environment due to the by-products and waste materials that are produced out of mixing these chemicals.

Precautionary Measures To Consider Before Taking Antibiotics

Synthetic antibiotics should really be given and used properly, responsibly and wisely. Although they are highly effective in treating bacterial infections, they must be taken with some precautionary measures in order to avoid their harmful side effects. First, you must give your body the chance to fight certain bacteria before turning to antibiotics for help because we have natural bacteria fighters. We must use our built in power to fight bacterial infections in order to make us less vulnerable to the bacteria that cause these infections. Second, make sure that you consult your physician first before taking any kind of antibiotic. You must know what kind of bacterial infection you have and you must not take other people's prescriptions. Third, you must finish taking your prescription up to the last day that the doctor instructed in order to prevent bacteria from becoming resistant to the drug and to prevent the formation of super bacteria. Lastly, you must consider your tolerance to a specific kind of antibiotic. Make sure that you will not have allergic reactions to these antibiotics and you must also make sure that you are not pregnant because some antibiotics can cause birth defects and spontaneous abortions.

Chapter 3
How You Can Benefit From Using Natural Antibiotics And Antivirals

The previous chapter discussed the advantages and disadvantages of synthetic antibiotics and it was pretty clear that they could cause a lot of side effects, most of which are detrimental to our health. So why, despite the clear evidence that the disadvantages outweigh the advantages of using synthetic antibiotics, do people still use them? The answer to that is maybe because people don't trust the natural counterparts of synthetic antibiotics enough to make them start using them as alternative. Another reason could be the high effectiveness of these synthetic antibiotics. Or maybe, most people don't realize that there are natural alternative in the first place. This chapter is dedicated to the discussion of the benefits that you can reap out of using natural antibiotics and antivirals.

Natural Antibiotics and Antivirals

Yes. Natural antibiotics and antivirals do exist. In fact, their effects are much more powerful than synthetic antimicrobials. Not only do they help boost our immune system and help us be more resistant to infections, they can also serve as anti-inflammatory agents that stop the spread of infection and cause acceleration of the healing process.

Synthetic Antimicrobials VS Natural Antivirals and Antibiotics

Both synthetic and natural antivirals and antibiotics are effective in treating infections caused by bacteria or viruses. The only difference between the two is that synthetic antivirals and antibiotics are produced from mixing different chemicals in a laboratory to achieve the antibacterial properties of natural antimicrobials. On the other hand, natural antivirals and antibiotics are not produced in a lab and they are just produced from natural ingredients, which are present in plants.

Although both synthetic and natural antivirals and antibiotics can kill good bacteria, natural antimicrobials lesser good bacteria compared to synthetic alternatives.

Benefits of Using Natural Antivirals and Antibiotics

So what really are the benefits of natural antimicrobials? There are absolutely many! For starters, we don't have to worry about harmful toxins that enter our systems together with the antibiotics or antivirals that we ingest since they are made from plants and herbs. Also, the usage of the natural cures makes it harder for bacteria and viruses to be resistant to them. This is because natural antivirals and antibiotics have more compounds unlike synthetically produced antimicrobials, which come from only one compound. These numerous compounds present in plants cooperate with one another in making a strong antibiotic or antiviral substance that most bacteria and viruses have difficulty surviving from. Most herbal remedies to bacterial and viral infections are also cheaper compared to artificially produced antibiotics and antivirals. Also, most of the herbs that have antiviral or antibacterial properties are common in our surroundings and are generally accessible to us. And the most important benefit that can be derived from the usage of natural antivirals and antibiotics is that they are do not pollute the environment. These herbal cures are not derived from harmful chemicals that can pollute the environment so using them is more environment-friendly.

There really are a lot of advantages in using natural remedies to viral and bacterial infections. Maybe it is time that we appreciate them and convert to the use of plant-based or natural cures.

Chapter 4
10 Herbs That Antibiotic Herbs And Antiviral Properties

The reason why natural remedies to infections caused by bacteria or viruses are called "natural" is because they can be found in nature or they are present in the environment. They can be in plants or herbs or fruits or even in leaves and flowers. Whoever or whatever made us really provided us with an environment filled with everything that we could possibly need.

Although there are only a relatively few number of herbs that have antimicrobial properties, they are widely common anyway so as far as accessibility to these plants is concerned, we don't have to really worry.

In this chapter, the top 10 herbs that have antiviral or antibacterial properties will be listed as well as the illnesses and diseases that they can cure (5 herbs with antibiotic properties and 5 herbs that have antiviral properties). Find out whether the herbs that will be mentioned are present in your garden or available in your local grocery store.

Herbs That Have Antibacterial Properties

Calendula
Calendulas are also commonly known as pot marigold. Its small yellow or orange flowers are used as first aid treatment to wounds and infections. It is also a known cure for acute conjunctivitis or what is commonly known as pinkeye.

Cinnamon

Little do we know, cinnamon have antibacterial properties too and they are not just added to our coffees or to our bread. Cinnamon has antibacterial properties that can help in digestion. It is also a carminative, which means that taking it can help us pass gas or fart.

Yarrow

Yarrows are plants that have leaves resembling that of the fern and have whitish or yellowish flowers. Its powdered flowers, when mixed with water, are used to treat sores. They can also quickly stop wounds from bleeding and can also cure urinary tract infections.

Garlic

Believe it or not, garlic is actually more effective in treating bacterial infections than Sir Fleming's Penicillin. It owes its antibacterial properties to the compound called allicin. Garlic is used as a first aid treatment for bites, especially dog bites and wounds.

Aloe vera

Aloe vera has been proven to be very effective in treating burns because of its cooling effect. It can also prevent infections and can boost up the healing of wounds.

Actually, aloe vera has an antiviral property too and it can even cure simple herpes virus.

Herb That Have Antiviral Properties

Ginger

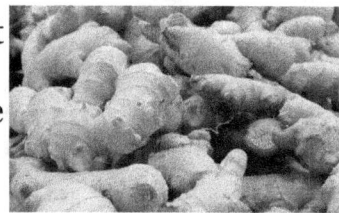

Ginger does not only make your voice sound more wonderful but it can also treat common colds and cough. Most people take ginger in the form of tea or in capsules.

Olive leaf

Colds, flu and herpes are just a few of the viral infections that the olive leaf can cure. It can be taken orally in the form of tea or capsules. Expectant mothers should avoid this though because it can cause problems for pregnant women.

Oregano

The oregano belongs to the family of mints and gives off an aromatic odour. It is commonly used as spice or seasoning but it has antiviral properties as well. It can cure cough and common colds.

Cranberry

Almost everyone loves cranberry juice. Well, you will love it more for sure because it has antiviral properties and it is an antioxidant as well. Cranberry is used as treatment for urinary tract infections and it also keeps your teeth strong by preventing the formation of plaque.

Cat's claw

Cat's claw helps us become less vulnerable to viral infections by making our immune system stronger. It also has anti-fungal and antibacterial properties.

Actually, there are a lot more plants and herbs that have antibacterial and/or antiviral properties. There are also some, which may have these properties, but there are no exhaustive studies about their healing potentials yet. Hopefully, more scientists and doctors would be more open to natural medicines and research on the medicinal properties of some plants in order to fully utilize nature's potential and in order for us to finally let go of these harmful synthetic medicines.

Chapter 5
Foods That Have Natural Antibiotic And Antiviral Properties

Of course, all of us need to eat. We get most of our energy, as well as the essential nutrients and vitamins that our body needs from the food that we eat. This is the reason why it is important that we should eat healthy as often as possible and try to munch on green leafy vegetables and juicy fruits instead of junk foods. Another thing that can entice us to eat healthy is the fact that there are certain foods that can help us build a stronger immune system and help us fight viral and bacterial infections.

This chapter will tell you what kinds of food have the ability to fight harmful bacteria and viruses and as you read along, try to see if you have been partaking these foods lately.

Fruits Rich In Vitamin C

Fruits such as pineapples, oranges, lemons, strawberries and watermelon, as well as vegetables such as cabbage, cauliflower and broccoli are rich in Vitamin C. This vitamin keeps our immune systems strong and formidable against the threat of viruses and bacteria, therefore making us more resistant to harmful infections.

Stocking up in Vitamin C and eating foods that contain Vitamin C can help treat and prevent colds, cough, and skin infections.

Coconut

Coconut is considered is the wonder fruit because all of its parts, from top to bottom, have a relevant use to humans. The coconut fruit has medium-chain fatty acids that can destroy pathogens and can battle viruses and bacteria.

Breast Milk

There is a reason why they say that breast milk is the best for babies. It is because it has antibacterial and antiviral properties aside from the nutrients, minerals and vitamins that it contains. The immune systems of babies are not as strong as those of older individuals. That is why mothers should breast feed their babies in order to give them the necessary protection from infectious diseases.

The human breast milk is an abundant source of lauric acid, a medium-chain fatty acid that kills pathogens. The lauric acid helps in building a strong immune system that is more resistant to viruses and bacteria. That is probably the reason why breast-fed babies don't get easily infected by viruses and bacteria compared to babies who are not.

Yoghurt

Yoghurts and other prebiotics increase the number of good bacteria living in the intestines and help make digestion smoother and easier. This ability helps fight off infections and helps kill the unwanted invaders such as viruses and bacteria.

So, are you keeping your immune system strong by eating these foods?

Chapter 6
Essential Oils That Have Natural Natural Antibiotic And Antiviral Properties

We are already done discussing herbs and foods that have natural antiviral and antibiotic properties. Now it is time to know what are some of the oils that possess these properties too.

This chapter will discuss the examples of oils that either have cure viral infections or bacterial infections or both. The specific diseases and ailments that they can cure will also be tackled.

Eucalyptus Essential Oil
Eucalyptus, which has insecticidal, antibacterial and anti-inflammatory properties, works perfectly with coughs, colds and irritations. It is even a common component or ingredient in rubs and ointments.

Lemon Essential Oil
Lemon is popularly known for its antiseptic powers. It can cleanse wounds and prevent infections as well as make the healing process faster. Lemon oil is also a good antioxidant and helps the body get rid of all the harmful toxins. It can also be used applied directly to the skin as cleanser or wound antiseptic.

Peppermint Essential Oil
Peppermint oil has a wide variety of uses. It can be utilized as an antiseptic, antibacterial, antiviral and even anti-inflammatory. It can cure nausea and other digestive diseases.

Virgin Coconut Oil

The oil extracted from heating coconut water can have a lot of health benefits. It can cure mouth infections, aid in digestion, heal wounds and help moisturise irritated and dry skin.

Chapter 7
How To Prepare And Use Natural Antibiotics And Antivirals

At this juncture, you are probably enticed already to use antivirals and antibiotics that can naturally be found around us because of the all the benefits of these natural antimicrobials and their availability. If you are, then it is very good that you are opening your doors for herbal cures and have considered using them instead of the artificially made ones.

So, the next question that you might be probably thinking in your head is this: *How do I use these natural antibiotics and antivirals?* The answer simple: Natural antivirals and antibiotics, just like synthetic antimicrobials, can be taken orally or topically, which means they can used as capsules, ointments, rubs or drinks. You have to be careful though and take into consideration a few things, just like your skin sensitivity to the natural treatment, which are applied topically. You must also take into consideration whether or not you are pregnant by the time you want to take these natural cured because some of them are harmful to pregnant women. You also need to have sufficient knowledge about the natural treatment that you want to take because some of them need to be diluted because of their strong effects.

This chapter will teach you to prepare natural antivirals and antibiotics as well as some natural remedies to skin problems and common ailments.

Preparing Natural Antivirals And Antibiotics

These are some simple and comprehensive steps on

how to prepare natural antimicrobials.

Natural Ointments

To make antiviral ointments, you must mix a maximum of 4 drops of one or more kinds of antiviral oils into pure coconut oil. Store in a glass jar and keep away from the reach of small children.

You can also utilize aloe vera as an ointment for burns. You can heat some aloe vera leaves before slicing them or you can slice them right away. Make sure that the transparent and colourless gooey substance is applied thoroughly throughout the burnt area.

Natural Cough Treatment

You can add mint leaves or Oregano leaves to a pot of water and boil it for 10 to 20 minutes. When the water becomes greenish or when you can already smell the aroma of the oregano or the mint mixing with the water, you can already drink the concoction.

Tea

To make your own herbal tea, all you need is a cup of hot water and any antiviral or antibacterial herb of your choice. Mix 1 tablespoon of antimicrobial herbs into a cup of water and leave it for 5 to 10 minutes. Drink when ready.

5 Natural Remedies For Common Ailments

It is important that we know some first aid treatments to common ailments because knowing such information can help save a life and prevent the spread of infection.

Here are 5 of the most common ailments acquired by most people and their natural cures:

- ## *Colds*

  - Getting a cold is one manifestation of a down immune system. It is your body's way of telling you that your natural defences are weak. The natural ways of treating colds is by drinking lots and lots of water to flush out the toxins in your body or by drinking tea with ginger. You can also opt for an overdose of Vitamin C through eating a lot of fruits rich in Vitamin C. You don't have to worry about overdosing in Vitamin C because it is water-soluble and the body can easily get rid of the excess Vitamin C in our system by urinating.

- ## *Stomach Ache*

- Drinking chamomile tea, or hot water with ginger or peppermint can help in indigestion.

- ## *Burns*

  - For less serious burns, aloe vera can be used to cool off the burnt part of the skin. If ever the burn is already a second-degree burn, go immediately to the hospital and let the doctors handle it.

- ## *Wounds*

  - A lot of herbs can be used to treat wounds and prevent them from getting infected. You can use lemon juice to clean the wound and to disinfect it. Warning: This may be very painful. You can also use garlic.

- ***Flu***
    - Flu can be cured by the help of drinking lots of fluids and eating vegetables and fruits rich in Vitamin C.

5 Natural Skin Care Remedies

Taking good care of the skin is important, especially for the ladies. Fortunately, nature also offers skin remedies that help us cure common skin diseases that are very stubborn to cure. There are also natural substances that can help us keep our skin young and glowing. Here are some of them:

1. ***Sugar mixed with coconut oil***
    - Does your skin appear dull and old? There's need to worry because nature has a solution for you! This mixture can serve as the natural equivalent to exfoliators. It can exfoliate dead skin cells, resulting to a softer and younger skin.

2. ***Vitamin C***
    - Vitamin C can help the skin become more youthful and firmer by facilitating the production of collagen to repair damaged cells.

3. ***Honey mixed with water***
    - This mixture serves as a perfect facial mask which helps cleanse the face and get rid of the excess oil and dirt.

4. _Coconut oil_

- Coconut oil can help moisturize the skin and keep it smooth and silky. It also prevents skin irritations.

5. _Oatmeal_

1. Oatmeal can actually cure inflammations and tone down allergic reactions.

Conclusion

I hope that you have learned a lot about antibiotics and antivirals after reading this ebook. I also hope that you will apply your new found knowledge about these antimicrobials into good use and encourage others to discover the benefits of natural treatments such as herbal antivirals and antibiotics, oils, as well as the foods that have and contain these antimicrobial properties.

I would also like to reiterate the importance of the body's natural defence system in fighting against these invaders. Just once again emphasize the point, we must allow our bodies to fight these harmful bacteria and viruses because we have the capability to do so. Impeding this capability can cause our defences to weaken and makes us more vulnerable to these invaders. Also, we must not allow ourselves to be dependent to these antimicrobials because some of them, especially the synthetically produced ones, have harmful effects on the body.

The herbs, oils, foods and other natural treatments stated in this ebook are not the only ones that have antimicrobial properties. There are still a lot of plants that also have the ability to fight off these harmful invaders waiting to be discovered.

Lastly, I hope that more researches about the effectiveness of natural treatment against harmful bacteria and viruses will surface in the medical industry and promote the use of natural cures instead of the synthetic ones.

From The Author

Thank you for taking the time to read this book. As an author, I understand the importance of creating books which my readers will find both enjoyable and informative. If you have the time and feel generous, please don't hesitate to leave an honest review of this book.........*Dr Alex Nelson·*

Other Books By Dr Alex Nelson

Cure Adrenal Fatigue Now!

Adrenal Fatigue doesn't have to control your life. There are solutions to help you diagnose and overcome this modern day symptom of stress. Within the pages of this easy-to-read guide, Dr. Alex Nelson, has outlined all the information you need to know in order to combat the dreaded listlessness and show you the steps necessary to recharge your energy levels and leave you feeling invigorated and ready for any of life's challenges. Discover the natural remedies that can and will change your life.

66

www.ingramcontent.com/pod-product-compliance
Lightning Source LLC
Chambersburg PA
CBHW070933180526
45168CB00003B/1053